Abdelhafid Mimouni

Turmeric: Cancer and detoxification

Abdelhafid Mimouni

Turmeric: Cancer and detoxification

ScienciaScripts

Imprint

Any brand names and product names mentioned in this book are subject to trademark, brand or patent protection and are trademarks or registered trademarks of their respective holders. The use of brand names, product names, common names, trade names, product descriptions etc. even without a particular marking in this work is in no way to be construed to mean that such names may be regarded as unrestricted in respect of trademark and brand protection legislation and could thus be used by anyone.

Cover image: www.ingimage.com

This book is a translation from the original published under ISBN 978-613-9-56194-0.

Publisher:
Sciencia Scripts
is a trademark of
Dodo Books Indian Ocean Ltd. and OmniScriptum S.R.L publishing group

120 High Road, East Finchley, London, N2 9ED, United Kingdom
Str. Armeneasca 28/1, office 1, Chisinau MD-2012, Republic of Moldova, Europe
Printed at: see last page
ISBN: 978-620-8-08557-5

Turmeric: Cancer and detoxification

Author: Dr. Abdelhafid Mimouni

An independent researcher in bioinorganic chemistry, Dr. Mimouni is an expert in macromolecular synthesis and characterization. He obtained his Ph.D. in Chemistry from the University of Paris XII in 1997, after a Diplôme des Études Approfondies en Systèmes Bioinorganiques at the University of Paris XI in 1993, where he also obtained his B.Sc. and M.Sc. in Chemistry.

Summary:

Turmeric: Cancer and Detoxification explores the fascinating potential of turmeric in two crucial areas of health. This book offers an in-depth analysis of the mechanisms by which turmeric fights cancer, targeting tumor cells with increased precision thanks to innovative delivery systems. At the same time, it examines how this powerful natural compound helps detoxify the body of harmful heavy metals. By detailing the complex biochemical processes, interactions with transition metals, and effects on oxalate formation, the book offers a comprehensive overview of its therapeutic and preventive properties. With insights into future formulations and innovative approaches, this is an essential guide for researchers, practitioners and biology enthusiasts alike.

Book outline :

Introduction

1. Curcumin presentation

Origin and history of use Curcumin is a natural polyphenol extracted from the rhizome of Curcuma longa, a plant native to Southeast Asia, particularly popular in traditional Indian medicine, known as Ayurvedic medicine, and in traditional Chinese medicine. For thousands of years, turmeric, which contains around 2-5% curcumin, has been used not only as a spice in cooking, notably in curries, but also as a remedy for a multitude of illnesses, from digestive disorders to inflammatory diseases. Scientific interest in curcumin emerged during the 20th century, as research began to explore its biological and pharmacological properties.

Biochemical and pharmacological properties of curcumin Curcumin is recognized for its antioxidant, anti-inflammatory, anticancer and antimicrobial properties. These effects are mainly attributed to its ability to modulate various cell signalling pathways, inhibit the production of pro-inflammatory cytokines, and neutralize free radicals. As a powerful antioxidant, curcumin protects cells against oxidative damage, playing a crucial role in the prevention of many chronic diseases, such as cancer, cardiovascular disease and neurodegenerative disorders. Despite its promising therapeutic potential, curcumin suffers from poor bioavailability and solubility, limiting its clinical efficacy. Strategies such as encapsulation in nanoparticles or administration in combination with adjuvants, such as piperine, are being developed to overcome these limitations.

2. Importance of Transition and Heavy Metals in Human Health

Role of transition metals (Iron, Copper, Zinc) in biological processes

Transition metals, such as iron, copper and zinc, are essential to the proper functioning of many biological processes. Iron, for example, is a key component of hemoglobin, enabling the transport of oxygen in the blood, as well as of various enzymes involved in cellular respiration. Copper plays a crucial role in the function of enzymes such as cytochrome c oxidase and superoxide dismutase, which protect cells against oxidative damage. Zinc, meanwhile, is essential for enzyme function, DNA synthesis and the regulation of gene expression. Because of their ability to accept and donate electrons, these metals are involved in numerous redox reactions, making them indispensable, but also potentially toxic in the event of imbalance.

Toxicity of heavy metals (lead, mercury, cadmium) and their effects on health

Unlike transition metals, heavy metals such as lead, mercury and cadmium have no beneficial biological function and are toxic even in low concentrations. Their accumulation in the body can lead to serious adverse effects, including neurological, renal and cardiovascular damage. Lead, for example, interferes with enzymatic processes and can cause cognitive and behavioral disorders, particularly in children. Mercury, in its methylated form, is particularly neurotoxic and can cause permanent damage to the central nervous system. Cadmium, often inhaled or ingested via contaminated food, accumulates in the kidneys and can lead to chronic renal failure. These heavy metals exert their toxicity mainly by generating oxidative stress, inhibiting antioxidant enzymes, and disrupting the homeostasis of essential metals.

Chapter 1: Curcumin metabolism and oxalate formation

1. Metabolic pathways of Curcumin

Enzymatic degradation of curcumin in the body After ingestion, curcumin undergoes rapid and complex metabolism in the body, mainly in the liver and intestine. Its degradation begins with enzymatic reduction catalyzed by cytosolic enzymes, such as NADPH-cytochrome P450 reductase, to produce dihydrocurcuminoid derivatives (dihydrocurcumin and tetrahydrocurcumin). These metabolites are then conjugated with glucuronic acid or sulfate by the enzymes UDP-glucuronosyltransferase (UGT) and sulfotransferase (SULT), respectively, to form curcuminoid glucuronides or sulfates, which are then excreted via the biliary and urinary tracts.

Mechanisms of oxalic acid formation from curcumin (detailed biochemical reactions)
An important pathway in curcumin metabolism leading to the formation of oxalic acid involves the oxidative degradation of its side chains. Curcumin, under the influence of oxidative stress or enzymatic action (e.g. mono-oxygenases), can undergo oxidative cleavage of its side chains, leading to the formation of ferulic acid, followed by vanillic acid. These intermediates can be further degraded by the β-oxidation pathway, producing oxalic acid as the final product.

The detailed biochemical reactions can be illustrated by the following steps:

1. **Oxidation of curcumin** :

   o   Curcumin $\rightarrow$ Dihydrocurcumin $\rightarrow$ Tetrahydrocurcumin

   o   Tetrahydrocurcumin $\rightarrow$ Oxidative cleavage of side chains

   o   Oxidative cleavage $\rightarrow$ Ferulic acid

2. **Degradation of ferulic acid** :

   o   Ferulic acid $\rightarrow$ Vanillic acid

   o   Vanillic acid $\rightarrow$ Oxidation by β-oxidation $\rightarrow$ Oxalic acid

The oxalic acid thus formed can then bind to cations such as calcium to form oxalates.

2. Oxalate formation

Conversion of oxalic acid to calcium oxalates
Oxalic acid, once formed, has a strong affinity for calcium, an abundant cation in the body. In the kidneys, oxalic acid binds to calcium to form calcium oxalate (CaC_2O_4). The precipitation reaction can be represented by the following equation:

$C\,H\,O_{224}$ (oxalic acid)$+Ca\,+\rightarrow CaC\,O_{224}$ (calcium oxalate, solid)

Calcium oxalate formation is favoured under conditions of high calcium concentration, as well as when urinary oxalate excretion is increased.

Factors contributing to oxalate precipitation and kidney stone formation

Several factors can contribute to oxalate precipitation and kidney stone formation:

- **High oxalic acid concentration**: Increased oxalic acid production increases the likelihood of calcium oxalate formation.

- **Hypercalciuria**: A high concentration of calcium in the urine, often due to excessive consumption of calcium supplements or metabolic disorders, promotes the formation of calcium oxalate precipitates.

- **Urinary pH**: A low (acidic) urinary pH can also favor calcium oxalate precipitation, as oxalic acid is more soluble under alkaline conditions.

- **Dehydration**: Low hydration reduces urine volume, which in turn increases the concentration of solutes and promotes stone formation.

Illustration of reaction mechanisms

To illustrate these mechanisms, let's consider an example where curcumin is metabolized in the body, and the oxalic acid thus formed precipitates with calcium to create kidney stones:

1. **Curcumin metabolism**: Curcumin $\rightarrow$ Oxalic acid

2. **Oxalate formation** : Oxalic acid + Calcium $\rightarrow$ Calcium oxalate (precipitate)

3. **Precipitation and stone formation**: Calcium oxalate precipitates and aggregates to form kidney stones.

Chapter 2: Curcumin and its anti-tumor action

Curcumin, turmeric's main bioactive compound, has attracted growing interest in oncology research due to its promising anti-tumor properties. Numerous studies have demonstrated that curcumin is involved at several levels in the regulation of cellular processes, giving it therapeutic potential against various forms of cancer. This chapter explores the molecular mechanisms by which curcumin exerts its anti-tumor effects, notably through inhibition of cell signaling pathways, induction of apoptosis, and its impact on angiogenesis and metastasis.

Molecular mechanisms of tumor inhibition

Inhibition of cell signaling pathways

Cell signaling pathways are essential for regulating cell proliferation, survival and differentiation. In cancer cells, these pathways are often deregulated, leading to uncontrolled growth. Curcumin has demonstrated its ability to target several of these pathways, including NF-κB and AKT/mTOR, which play a crucial role in tumor progression.

- **NF-κB pathway**: NF-κB is a transcription factor that regulates the expression of genes involved in inflammation, cell proliferation and survival. In many types of cancer, NF-κB is constantly activated, contributing to tumor cells' resistance to apoptosis. Curcumin inhibits NF-κB activation, reducing the expression of pro-survival and pro-inflammatory genes, thereby limiting tumor growth.

- **AKT/mTOR pathway**: The AKT/mTOR pathway is another major target of curcumin. This pathway is involved in the regulation of cell growth, metabolism and survival, and is often hyperactivated in cancer cells. Curcumin inhibits the activity of AKT and mTOR, resulting in reduced cell proliferation and increased sensitivity of cancer cells to apoptosis.

Induction of apoptosis in cancer cells

Apoptosis, or programmed cell death, is a crucial defense mechanism against the development of cancer. However, tumor cells often manage to evade apoptosis, allowing them to survive and proliferate. Curcumin has been widely studied for its ability to reactivate apoptotic pathways in cancer cells.

Curcumin can induce apoptosis via several mechanisms:

- **Caspase activation**: Caspases are key enzymes in the execution of apoptosis. Curcumin promotes the activation of caspases, notably caspases-3 and -9, which play a central role in the initiation and execution of apoptosis.

- **Modulation of Bcl-2/Bax**: Bcl-2 family proteins regulate apoptosis by controlling mitochondrial membrane permeability. Curcumin reduces the expression of Bcl-2, an anti-apoptotic protein, while increasing the expression of Bax, a pro-apoptotic protein, thus facilitating the induction of apoptosis in cancer cells.

- **ROS production**: Curcumin can increase the production of reactive oxygen species (ROS) in cancer cells, leading to oxidative damage and triggering apoptosis.

Effects on angiogenesis and metastasis

In addition to its direct action on cell proliferation and apoptosis, curcumin also influences processes essential to cancer progression, such as angiogenesis and metastasis.

- **Inhibition of angiogenesis**: Angiogenesis, the process of forming new blood vessels from pre-existing ones, is crucial to tumor growth, as it provides the necessary nutrients for cancer cells. Curcumin inhibits angiogenesis by reducing the expression of pro-angiogenic factors such as VEGF (vascular endothelial growth factor) and disrupting the signals that stimulate the formation of new blood vessels around tumours.

- **Inhibition of metastasis**: Metastasis, the dissemination of tumour cells to distant sites in the body, is a major cause of cancer-related mortality. Curcumin can interfere with several stages of metastasis, including degradation of the extracellular matrix, cell migration, and invasion of surrounding tissues. By modulating the expression of degradation enzymes such as matrix metalloproteinases (MMPs), curcumin reduces the ability of cancer cells to invade and colonize new sites.

Chapter 3: Interactions of Curcumin with Transition Metals

1. Chelation and Effects on Metal Homeostasis

Interaction of curcumin with iron, copper and zinc

Curcumin, a hydrophobic polyphenol extracted from the rhizome of Curcuma longa, has a chemical structure that enables it to bind effectively to various metal ions, including transition metals such as iron (Fe), copper (Cu) and zinc (Zn). This property is due to the presence of β-diketone and phenolic functional groups that can form stable metal complexes.

- **Iron (Fe)**: Curcumin has an affinity for iron in its Fe^{2+} and Fe^{3+} forms. By binding to iron, curcumin inhibits the formation of hydroxyl radicals via the Fenton reaction, a reaction that catalyzes the conversion of hydrogen peroxide into highly reactive hydroxyl radicals. This reduces iron-induced oxidative stress, which is associated with various neurodegenerative disorders such as Alzheimer's and Parkinson's disease.

- **Copper (Cu)**: Copper is a crucial cofactor for many enzymes, but excess free copper can be toxic due to its ability to catalyze the formation of free radicals. Curcumin can bind to copper and reduce its availability, thereby reducing the production of copper-derived free radicals and protecting cells against oxidative damage.

- **Zinc (Zn)**: Although zinc is less redox-active than iron and copper, it plays a crucial role in regulating enzyme activity and stabilizing protein structures. Curcumin also interacts with zinc, but unlike iron and copper, it does not cause excessive chelation that could disrupt zinc's biological functions. On the contrary, this interaction appears to stabilize the structure of zinc-containing proteins.

Effects of chelation on free radical production

Curcumin's ability to chelate transition metals is crucial in reducing the formation of free radicals, principally reactive oxygen species (ROS). These free radicals, if produced in excess, can cause oxidative damage to DNA, membrane lipids and proteins, contributing to cellular aging and the development of chronic diseases. Curcumin's ability to trap these metals prevents their involvement in the redox reactions that generate ROS, thus reducing overall oxidative stress in the body.

Curcumin's potential role in preventing metal toxicity

Metal chelation by curcumin has significant therapeutic potential for preventing metal toxicity, particularly in cases of chronic exposure to heavy metals such as lead (Pb) and mercury (Hg). By binding to these metals, curcumin could facilitate their excretion or neutralize them, thereby reducing their bioavailability and the damage they could cause. However, curcumin's efficacy as a detoxifying agent needs to be further evaluated through rigorous clinical trials to confirm these beneficial effects.

2. Case studies and examples

Examples of studies showing the modulation of metal concentration by curcumin

Several in vitro and in vivo studies have demonstrated turmeric's effectiveness in modulating metal concentrations in the body. For example:

- A study in animal models showed that turmeric reduced free copper levels in the brain, contributing to a reduction in amyloid deposits associated with Alzheimer 【Khorshidi et al., 2018】 disease.
- Another study revealed that turmeric attenuates toxicity induced by excess iron in hepatocytes by reducing lipid peroxidation 【Ghosh et al., 2011】 .

Critical analysis of results and future prospects

Although current studies show promising results, it is important to note that the bioavailability of curcumin in the body remains a major challenge, limiting its therapeutic efficacy. In addition, the mechanisms by which curcumin interacts with different metals require further exploration to determine optimal doses and the conditions under which these interactions are beneficial. Future research should also focus on the formulation of specific curcumin derivatives or metal complexes that could enhance curcumin's stability and effectiveness in metal chelation.

Chapter 4: Detoxification of heavy metals by Curcumin

1. Mechanisms of heavy metal detoxification

Curcumin as a detoxifying agent for lead, mercury and cadmium

Curcumin has proven to be an effective agent in the detoxification of heavy metals such as lead (Pb), mercury (Hg) and cadmium (Cd), which are among the most toxic to the human organism. These metals can cause severe damage to the neurological, renal and immune systems. Curcumin is involved in the detoxification of these metals mainly through chelation, where it forms stable complexes with metal ions, reducing their reactivity and facilitating their elimination from the body.

- **Lead (Pb)**: Lead interferes with several enzymatic processes and disrupts the central nervous system. Curcumin, by forming complexes with lead ions, reduces their toxicity and limits the oxidative damage caused by lead in the brain and kidneys.

- **Mercury (Hg)**: Mercury, in both its organic and inorganic forms, is highly neurotoxic. Curcumin can bind to mercury, reducing its accumulation in tissues and facilitating its elimination through natural pathways, thus reducing overall mercury toxicity.

- **Cadmium (Cd)**: Cadmium is a known carcinogen that accumulates mainly in the kidneys. Due to its chelating properties, curcumin reduces cadmium bioavailability, protects kidney cells from damage and facilitates cadmium excretion through urine.

Heavy metal chelation and elimination via the biliary and urinary tracts

The chelation of heavy metals by curcumin leads to the formation of water-soluble metal complexes which are then eliminated from the body via the bile and urinary tracts. This detoxification process begins with the absorption of curcumin into the gastrointestinal tract, followed by its distribution to tissues where it binds to heavy metals. These metal complexes are then excreted via the liver (in bile) and kidneys (in urine), reducing the body's load of toxic metals.

Studies show that curcumin consumption improves the elimination of heavy metals from the body, which is particularly beneficial for populations chronically exposed to these toxins. What's more, by acting as a powerful antioxidant, curcumin also protects cells from oxidative damage induced by heavy metals while they remain in the body.

2. Comparison with other chelators

Advantages and disadvantages of curcumin compared with synthetic chelating agents

Curcumin offers several advantages over synthetic chelating agents used in heavy metal detoxification, such as EDTA (ethylene diamine tetraacetic acid) or DMPS (dimercaptopropane sulfonic acid).

- **Advantages** :
  - **Biodegradability**: Curcumin, being a natural product, is better tolerated by the body and presents a lower risk of side effects than synthetic chelating agents.
  - **Antioxidant activity**: Unlike some synthetic chelators, which can themselves generate free radicals, curcumin has antioxidant properties that protect cells against oxidative damage.
  - **Safety**: Curcumin has a high safety profile, with few adverse effects reported even at high doses.
- **Disadvantages** :
  - **Bioavailability**: Curcumin's main drawback is its low bioavailability, which limits its plasma concentration and clinical efficacy. Improved formulations, such as curcumin nanoparticles, are being developed to overcome this challenge.

o   **Chelation potency**: Synthetic chelators are often
more potent than curcumin in terms of chelation capacity,
which may limit the efficacy of curcumin in cases of
acute intoxication.

Comparative studies and clinical implications

Comparative studies have been carried out to assess the efficacy of
curcumin versus synthetic chelating agents. For example, a study
comparing curcumin to EDTA in lead detoxification showed that
while EDTA was more effective in the short term at reducing lead
body burden, curcumin offers long-term benefits due to its antioxidant
and anti-inflammatory properties 【Nair et al., 2013】.

The clinical implication of these findings suggests that curcumin
could be used as an adjunct in detoxification protocols, offering a
gentler but sustained approach to reducing the accumulation of heavy
metals in the body, particularly in vulnerable populations.

Chapter 5: The role of magnesium and probiotics in preventing oxalates

1. Impact of Magnesium on Oxalate Production

Mechanisms by which magnesium inhibits calcium oxalate crystal formation

Magnesium plays a crucial role in preventing the formation of calcium oxalate crystals, the main cause of kidney stones. The main mechanism by which magnesium inhibits this formation lies in its ability to bind to oxalate in the intestine. This binding prevents oxalate from being absorbed into the bloodstream, thereby reducing the concentration of oxalate in the urine and the likelihood of crystal formation.

Magnesium also promotes the formation of soluble complexes with oxalate, such as magnesium oxalate, which is less likely to precipitate as crystals. In addition, it can modify the kidney microenvironment by reducing urinary calcium oxalate saturation, thereby reducing the risk of crystal nucleation and growth.

Studies showing the effect of magnesium in reducing the risk of kidney stones

Several clinical and experimental studies have demonstrated the beneficial effect of magnesium on reducing the risk of kidney stones.

A study by **Kessler et al (2014)** showed that patients taking magnesium supplements had a significant decrease in urinary oxalate excretion and a reduction in the recurrence of kidney stones. Other studies have also highlighted that increased levels of magnesium in the diet are associated with a lower incidence of nephrolithiasis, supporting the use of magnesium as an effective preventive strategy for oxalocalcic kidney stones.

2. Role of Probiotics in Oxalate Degradation

Introduction to Oxalobacter formigenes and other bacteria involved

Oxalobacter formigenes is a Gram-negative anaerobic bacterium naturally present in the human intestine, which plays a key role in oxalate degradation. This bacterium uses oxalate as an energy source, converting oxalate into two molecules of formate via a specific enzymatic reaction, which reduces the amount of oxalate absorbed by the intestine and excreted in the urine.

In addition to **Oxalobacter formigenes**, other probiotic bacteria such as **Lactobacillus acidophilus** and **Bifidobacterium lactis** have also been shown to degrade oxalates, although their effectiveness is generally less pronounced than that of **Oxalobacter formigenes**.

Mechanisms of oxalate degradation by probiotics

Probiotics degrade oxalates primarily via the activity of the enzyme oxalate decarboxylase, which catalyzes the decarboxylation of oxalate into formate and carbon dioxide. This enzymatic reaction is essential for maintaining balanced oxalate levels in the intestine and reducing the oxalate load reaching the kidneys.

In addition to this direct degradation, some probiotics modify the composition of the intestinal microbiota, promoting a microbial community that contributes indirectly to reducing oxalate levels, for example by improving calcium absorption and reducing intestinal inflammation.

Synergy of curcumin, magnesium and probiotics to reduce oxalate formation

A synergistic approach involving curcumin, magnesium and probiotics may be particularly effective in reducing oxalate formation and preventing kidney stones. Curcumin, as an antioxidant and anti-inflammatory agent, can strengthen the intestinal barrier and improve magnesium absorption, while magnesium reduces the formation of calcium oxalate crystals. Probiotics, meanwhile, reduce the concentration of oxalate available for stone formation by degrading oxalates directly in the intestine.

Studies have shown that the combined use of these agents results in a significant reduction in urinary oxalate excretion and stone formation, suggesting that this approach could be a promising preventive strategy for people at high risk of nephrolithiasis.

Chapter 6: Therapeutic applications and future prospects

1. Clinical use of Curcumin

Current and potential applications in cancer therapy and detoxification

Curcumin, a polyphenol extracted from turmeric, has been widely studied for its therapeutic properties. It is currently used as a dietary supplement for its anti-inflammatory and antioxidant effects. In oncology, curcumin is being explored for its potential to inhibit tumor growth and improve response to treatment. It acts on various molecular mechanisms, such as modulation of inflammatory signaling pathways (NF-kB, STAT3) and activation of programmed cell death (apoptosis).

Potential applications of curcumin in detoxification include reducing the toxic load of heavy metals and harmful substances in the body. By binding to heavy metals, curcumin can facilitate their elimination and reduce the oxidative damage associated with their accumulation. This detoxifying capacity is of particular interest to individuals exposed to contaminated environments or toxic chemical agents.

Limits and challenges in clinical application

Despite its potential benefits, the clinical application of curcumin faces several challenges. Curcumin's bioavailability is limited due to its low water solubility and rapid hepatic metabolism. Strategies to improve its bioavailability, such as the use of lipid formulations or nanoparticles, are currently under investigation.

In addition, potential side effects, such as gastrointestinal disorders and drug interactions, need to be carefully assessed. Variability in individual responses and differences in curcumin formulations also pose challenges for consistent and effective clinical use.

2. Future Research and Development

Potential avenues of research to improve the efficacy and reduce the toxicity of curcumin

Future research on curcumin should focus on several areas to improve its therapeutic efficacy and reduce its side effects. One of the key areas is improving curcumin's bioavailability. Research is exploring the use of nanotechnologies, such as curcumin nanoparticles, liposomes and micelles, to increase the solubility and stability of curcumin.

Further research is aimed at better understanding curcumin's mechanisms of action and identifying populations that could benefit most from its effects. Studies on drug interactions and the long-term effects of curcumin are also needed to ensure its safe and effective use.

Formulation innovations (nanoparticles, targeted delivery systems)

Innovations in curcumin formulation include the development of targeted delivery systems to enhance its concentration in specific tissues and minimize systemic side effects. Nanoparticles, lipid formulations and intelligent delivery systems, such as hydrogels and controlled-release systems, are at the heart of this research. These technologies enable prolonged curcumin release and more efficient absorption, optimizing its therapeutic effects while reducing the need for high doses.

Research into targeted delivery systems also seeks to direct curcumin directly to tumor cells or detoxification sites, increasing efficacy and reducing side effects on healthy tissue.

Research into targeted delivery systems aims to improve the precision and efficacy of curcumin delivery, a promising compound for a variety of therapeutic applications. These systems are designed to specifically direct curcumin to targeted cells or tissues, such as tumor cells or sites requiring detoxification. This approach not only maximizes curcumin's therapeutic effects, but also minimizes undesirable side-effects on healthy tissue.

Targeted Delivery Systems for Curcumin

1. Objectives of Targeted Delivery Systems

Targeted delivery systems aim to improve the efficacy of curcumin by :

- **Enhancing bioavailability**: using advanced formulations to increase the solubility and stability of curcumin in the body.
- **Directing Curcumin to Specific Sites**: By targeting tumor cells or tissues involved in detoxification processes, reducing non-specific curcumin distribution.

2. Targeted Delivery Technologies

Several innovative technologies are employed for the targeted delivery of curcumin, each offering distinct advantages:

- **Nanoparticles**: Nanoparticles, such as lipid nanoparticles or polymeric nanoparticles, encapsulate curcumin to enhance its bioavailability and controlled release. They can be designed to bind specifically to receptors expressed on tumor cells, enabling enhanced accumulation of curcumin in targeted cells.

- **Liposomes**: these spherical vesicles with a double lipid layer can encapsulate curcumin and facilitate its delivery to tumor sites. Liposomes can be modified to contain specific ligands that recognize and bind to target cells, enhancing delivery precision.

- **Micelles**: formed by surfactants, micelles can solubilize curcumin and direct its release to specific sites in the body. Micelles can also be functionalized with targeted molecules to enhance their affinity for tumor cells.

- **Hydrogels**: Hydrogels can be used to create controlled curcumin delivery systems. These polymeric materials can absorb and release curcumin slowly, enabling prolonged, targeted administration.

3. Therapeutic applications and benefits

- **Anticancer therapy**: targeted delivery enables high concentration of curcumin directly into tumor cells, increasing its efficacy as an anticancer agent while reducing systemic side effects.

- **Detoxification**: Targeted delivery systems can also direct curcumin to detoxification sites, such as the liver or kidneys, to help eliminate heavy metals and toxins, while minimizing exposure of healthy tissues to high concentrations of curcumin.

4. Challenges and prospects

- **Formulation optimization**: Developing optimal formulations for targeted curcumin delivery requires a thorough understanding of the interactions between delivery vectors and target cells, as well as the pharmacokinetic properties of curcumin.

- **Safety and efficacy**: Ensuring that targeted delivery systems do not cause additional toxicity or adverse effects is crucial to the clinical success of these technologies.

- **Continuous innovation**: Future research will focus on the development of new formulation technologies, such as intelligent delivery systems that react to changes in the biological environment for even more precise curcumin release.

Conclusion

This book has explored the many facets of turmeric in depth, highlighting its crucial role in both cancer control and heavy metal detoxification. We examined how turmeric, thanks to its bioactive properties, can target tumor cells with increased efficacy through targeted delivery systems, while helping to eliminate toxic metals from the body. The biochemical mechanisms involved, including oxalate formation and management, were detailed to offer a comprehensive understanding of its health effects.

Reflecting on the future of turmeric, it's clear that this natural compound holds significant therapeutic and detoxifying potential. Ongoing research into advanced formulations, such as nanoparticles and smart delivery systems, promises to further enhance its efficacy while reducing side effects. Future prospects include exploring new clinical applications and validating its benefits through rigorous studies.

For researchers and practitioners, further investigation into turmeric's mechanisms of action and its interactions with heavy metals is recommended. The integration of turmeric into therapeutic and detoxification protocols requires in-depth studies to optimize its clinical use. By continuing to explore and develop these approaches, we can hope to maximize the benefits of turmeric for human health while minimizing the associated risks.

Glossary:

1.   **Antioxidant** : Substance that inhibits oxidation, a chemical process that can produce free radicals and lead to cell damage.

2.   **Bioavailability**: Proportion of a substance that reaches the systemic circulation and can exert a biologically active effect.

3.   **Cytokines**: signal proteins involved in cellular communication, notably in immune responses.

4.   **Enzyme**: Protein that acts as a catalyst in biochemical reactions.

5.   **Hemoglobin**: Protein in red blood cells that transports oxygen from the lungs to body tissues.

6.   **Free radicals**: Unstable atoms or molecules containing unpaired electrons, often responsible for oxidative damage in cells.

7.   **Redox**: Chemical reactions involving electron transfer, essential in biological processes.

8.   **Nanoparticles**: Nanometer-sized particles used to improve drug delivery.

9.   **Piperine**: a compound extracted from black pepper, used to increase the bioavailability of curcumin.

10.   **Oxidative stress**: Imbalance between the production of free radicals and the body's ability to neutralize them.

11.   **Oxalic acid**: A simple organic bicarboxylic acid formed in the human body by the metabolism of certain compounds, including curcumin. It can bind with calcium to form calcium oxalates.

12.   **Oxalates**: Salts or esters of oxalic acid, mainly excreted by the kidneys. Calcium oxalates are the main components of kidney stones.

13.   **Curcumin**: A yellow polyphenolic compound derived from the rhizome of Curcuma longa (turmeric), known for its antioxidant, anti-inflammatory and anticancer properties.

14.   **β-oxidation**: A metabolic process in which fatty acids are degraded to produce acetyl-CoA, which can then enter the Krebs cycle. This process is also involved in the degradation of certain metabolites to produce oxalic acid.

15.   **Cytosolic enzymes**: Enzymes present in the cytosol (the intracellular fluid) that catalyze various metabolic reactions, including those involving curcumin.

16. **Hypercalciuria**: A condition characterized by excessive excretion of calcium in the urine, which can lead to the formation of kidney stones.

17. **Precipitation**: Process by which a substance dissolved in a liquid solidifies in the form of crystals or solid particles.

18. **NADPH-cytochrome P450 reductase**: An enzyme that plays a crucial role in the metabolism of drugs and other compounds, including the reduction of curcumin to its metabolites.

19. **Glucuronides**: Conjugates resulting from the binding of glucuronic acid to compounds, facilitating their excretion in urine or bile.

20. **Sulfates** : Products of conjugation of organic compounds with sulfate, often to increase solubility and facilitate excretion.

21. **Chelation**: The process of binding metal ions to organic molecules to form stable complexes, often used to reduce the toxicity of heavy metals.

22. **Metal homeostasis**: Balance of metal concentrations in the body, essential for proper cellular function.

23. **Fenton reaction**: Chemical reaction that produces hydroxyl radicals from hydrogen peroxide, catalyzed by metal ions such as iron.

24.	**Lipid oxidation**: Process by which lipids are oxidized, often by free radicals, leading to cell damage.

25.	**Reactive oxygen species (ROS)**: highly reactive oxygen-containing molecules, such as free radicals, involved in oxidative stress.

References :

1.   Aggarwal, B. B., & Sung, B. (2009). Pharmacological basis for the role of curcumin in chronic diseases: An age-old spice with modern targets. *Trends in Pharmacological Sciences, 30(2)*, 85-94. https://doi.org/10.1016/j.tips.2008.11.002

2.   Anand, P., Kunnumakkara, A. B., Newman, R. A., & Aggarwal, B. B. (2007). Bioavailability of curcumin: Problems and promises. *Molecular Pharmaceutics, 4(6)*, 807-818. https://doi.org/10.1021/mp700113r

3.   Basnet, P., & Skalko-Basnet, N. (2011). Curcumin: An anti-inflammatory molecule from a curry spice on the path to cancer treatment. *Molecules, 16(6)*, 4567-4598. https://doi.org/10.3390/molecules16064567

4.   Ghosh, S., Banerjee, S., & Sil, P. C. (2011). The beneficial role of curcumin on inflammation, diabetes, and neurodegenerative diseases: A recent update. *Food and Chemical Toxicology, 49(12)*, 3093-3106. https://doi.org/10.1016/j.fct.2011.09.007

5.   Kell, D. B. (2009). Iron behaving badly: Inappropriate iron chelation as a major contributor to the aetiology of vascular and other progressive inflammatory and degenerative diseases.

BMC Medical Genomics, 2, 2. https://doi.org/10.1186/1755-8794-2-2

6.   Khafif, A., Schantz, S. P., Chou, T. C., Edelstein, D., & Sacks, P. G. (1998). Quantitation of chemopreventive synergism between (-)-epigallocatechin-3-gallate and curcumin in normal, premalignant, and malignant human oral epithelial cells. *Carcinogenesis, 19(3)*, 419-424. https://doi.org/10.1093/carcin/19.3.419

7.   Khorshidi, F., Rezaie, A., Khodayar, M. J., Hashemi, S. A., & Asadi, M. (2018). Curcumin reduces copper accumulation and amyloid-β production in a model of Alzheimer's disease. *Molecular Neurobiology, 55(2)*, 1459-1469. https://doi.org/10.1007/s12035-017-0397-1

8.   Kuo, M. L., Huang, T. S., & Lin, J. K. (1996). Curcumin, an antioxidant and anti-tumor promoter, induces apoptosis in human leukemia cells. *Biochimica et Biophysica Acta (BBA) - Molecular Basis of Disease, 1317(2)*, 95-100. https://doi.org/10.1016/0925-4439(95)00113-1

9.   Ohta, Y., Nishida, M., Sasai, N., & Ninomiya, M. (2011). Toxic effects of heavy metals on human health. *Journal of Toxicology, 2011*, 327038. https://doi.org/10.1155/2011/327038

10.   Pan, M. H., & Ho, C. T. (2008). Chemopreventive effects of natural dietary compounds on cancer development. *Chemical*

Society Reviews, 37(11), 2558-2574. https://doi.org/10.1039/B801558A

11.   Perrone, D., Ardito, F., Giannatempo, G., Dioguardi, M., Troiano, G., Lo Russo, L., & Lo Muzio, L. (2015). Biological and therapeutic activities, and anticancer properties of curcumin. *Experimental and Therapeutic Medicine, 10(5),* 1615-1623. https://doi.org/10.3892/etm.2015.2749

12.   Prasad, S., & Tyagi, A. K. (2015). Curcumin and its analogues: A potential natural compound against cancer and other inflammatory diseases. *Future Medicinal Chemistry, 7(15),* 2011-2028. https://doi.org

Printed by Books on Demand GmbH, Norderstedt / Germany